The Changing Profile of Corporate Climate Change Risk

Dr Mark C. Trexler
Director, Climate Risk, DNV-KEMA Energy & Sustainability,
Portland, OR, USA

Laura H. Kosloff
Attorney at Law, Portland, OR, USA

First published in 2012 by Dō Sustainability

87 Lonsdale Road, Oxford OX2 7ET, UK

ISBN 978-1-909293-01-4 (eBook-ePub)
ISBN 978-1-909293-02-1 (eBook-PDF)
ISBN 978-1-909293-00-7 (Paperback)

A catalogue record for this title is available from the British Library. Library of Congress Cataloging in Publication data applied for.

At Dō Sustainability we strive to minimize our environmental impacts and carbon footprint through reducing waste, recycling and offsetting our CO_2 emissions, including those created through publication of this book. For more information on our environmental policy see **www.dosustainability.com**.

Page design and typesetting by Alison Rayner
Cover by Becky Chilcott

For further information on Dō Sustainability, visit our website: **www.dosustainability.com**

DōShorts

Dō Sustainability is the publisher of **DōShorts**: short, high-value ebooks that distil sustainability best practice for busy professionals. Each DōShort addresses one sustainability challenge at a time and can be read in 90 minutes.

Be the first to hear about new DōShorts

Our monthly newsletter includes links to all new and forthcoming DōShorts. We also link to a free extract from each new DōShort in our newsletter, and to blogs by our expert authors. Sign up for the newsletter at **www.dosustainability.com**.

Recently published and forthcoming titles you'll hear about include:

- *Promoting Sustainable Behaviour: A Practical Guide to What Works*
- *Sustainable Transport Fuels Business Briefing*
- *How to Make your Company a Recognised Sustainability Champion*
- *Product Sustainability*
- *Corporate Climate Change Adaptation*
- *Making the Most of Standards*
- *Sustainability Reporting for Small and Medium-sized Enterprises*
- *The Changing Profile of Corporate Climate Change Risk*
- *Solar Photovoltaics: Business Briefing*
- *Green Jujitsu: The Smart Way to Embed Sustainability into Your Organisation*
- *The First 100 Days on the Job: How to Plan, Prioritise and Build a Sustainable Organisation*

Write for us, or suggest a DōShort

Please visit **www.dosustainability.com** for our full publishing programme. If you don't find what you need, write for us! Or Suggest a DōShort on our website.

We look forward to hearing from you!

Abstract

BUSINESS RISK associated with climate change is commonly assumed to be primarily policy driven. Many companies internalize the current stalemate over global climate policy into a perception that climate risk is no longer a critical issue. Business climate risks, however, include operational and supply chain (physical) risk, brand risk, market-driven structural risk, and liability risk. As national and global policy to materially reduce climate change is delayed, it is business-prudent to assume that the level of climate risk is increasing. Even if policy risk might seem lower today than a few years ago, political will can change quickly. Should physical impacts of climate change manifest in dramatic ways, for example, draconian climate policy is likely to follow quickly. Companies are well served to rethink their perceptions of climate risk in today's changing risk environment and evaluate whether they are effectively positioned.

Contents

CONTENTS

The Authors

DR MARK C. TREXLER has 25 years of experience with business climate change risk management, having joined the Climate, Energy and Pollution Program of the World Resources Institute in Washington, DC in 1988. Mark founded Trexler Climate + Energy Services (TC+ES) in 1991, pioneering corporate climate change strategy work, including the first GHG inventories, the first carbon offsets, the first corporate risk management strategies, and the first climate-neutral products and services. Mark has served as a lead author on climate change mitigation for the Intergovernmental Panel on Climate Change. Mark joined Det Norske Veritas (DNV) in 2009.

LAURA KOSLOFF has worked as an environmental attorney since 1985, including for the Environmental Law Institute in Washington, DC as a research attorney and as Associate Editor for the *Environmental Law Reporter*, as trial attorney for the US Department of Justice, as General Counsel to Trexler Climate + Energy Services (TC+ES), and as Associate General Counsel to EcoSecurities Group plc. At EcoSecurities, she supported a team of more than 20 climate change and project experts and developed climate-related contracts. As General Counsel of TC+ES, she participated in GHG market forecasting, carbon project development, and management of an emerging entrepreneurial enterprise.

CHAPTER 1

Executive Summary

CLIMATE CHANGE HAS BEEN CHARACTERIZED as a business risk, at least for major greenhouse gas (GHG) emitters, for more than 20 years.[1] Today, climate change and related risk variables (e.g. water scarcity and extreme events) increasingly rank toward the top of the business risk list published annually by the World Economic Forum.[2]

Climate change can translate into business risks in a number of ways:

- *Physical risk*, including direct impacts of climate change on a company's operations, supply chains, and financial performance;

- *Brand risk*, including the impact of consumer and stakeholder perceptions on corporate competitiveness;

- *Policy risk*, including the impacts of climate change policy and regulatory mandates on a company's operations, supply chains, and competitive advantage;

- *Structural risk*, including the impacts of climate change-influenced market forces on the supply of and demand for a company's products and services; and

- *Liability risk*, including litigation or legislation that could assign corporate liability for GHG emissions, potentially retroactively.

Several of these risks can manifest themselves at both ends of the corporate risk management time-line, i.e. through measures undertaken too early and too aggressively, or too little and too late. Stakeholders routinely pressure companies in virtually every business sector to take corporate action on climate change, even before any regulatory regime is in place. Yet it is a rare investor indeed who willingly and explicitly accepts reduced corporate performance in the near term as a tradeoff for being better hedged against future climate risks. Corporate first-movers undertaking aggressive mitigation strategies can find themselves under pressure from shareholders due to short-term impacts on the bottom line, face competitive disadvantage if competitors do not pursue the same initiatives, and even incur brand risk for their efforts if public attention to those efforts leads to 'greenwashing' charges.

Companies therefore walk a fine line when it comes to managing potential climate risks – assuming they are actually aware of the risks. Playing it safe, much of the corporate action on climate change has taken the form of low-risk voluntary measures, including harvesting low-hanging fruit (e.g. energy efficiency). While that can lead to significant cost savings, it usually runs out long before a company is able to accomplish serious emissions reductions. More importantly, it addresses just one of the potential risks companies face from climate change.

For companies wanting to undertake more material risk management efforts, whether mitigation or adaptation-based, physical and policy uncertainties surrounding future climate change and climate change policy are a major challenge. Corporate efforts can be frustrated by societal risk management inaction on the one hand (resulting in delayed policies and more climate change than the company might have anticipated) or societal risk management actions that are more aggressive than the company might have anticipated. The characteristics of climate change as a risk problem put companies at a significant risk management disadvantage, even as those risks grow.

An obvious question for business observers is just how sure scientists are about climate change and the existence of physical climate risks. There is no question that consensus exists among the scientific community that anthropogenic climate change poses serious risks, notwithstanding a range of continuing uncertainties regarding the magnitude, pace, and impacts of climate change. In interpreting the continuing scientific debates over these aspects of climate change, we have to remember that there's virtually nothing that scientists agree upon universally. This should not be interpreted as somehow undermining scientific certainty about climate change; there's almost nothing that scientists consider 'certain' in the way the term is commonly used.

From a business risk standpoint, it is useful to characterize a range of climate change scenarios against which potential business risks can be compared and evaluated. Five scenarios are profiled below, representing a large part of the potential distribution of climate change and climate policy outcomes:

- *Scenario 1: Issue collapse.* The prospects of climate change, and the pressure for policy action on climate change, could come to an end.

- *Scenario 2: Stay the policy course.* This scenario can be thought of as reflecting a continuation of current policies, and comparable to an explicit or implicit carbon price of US\$5–30/ton of CO_2 equivalent.

- *Scenario 3: Policy induced atmospheric stabilization of CO_2.* This scenario is based on emergence of the political will to pursue the aggressive emissions reductions and technology development initiatives that would be necessary to stabilize GHG concentrations in the atmosphere.

- *Scenario 4: Policy induced return to 350 ppm CO_2.* This scenario carries Scenario 3 further by suggesting an actual reversal in the accumulation of GHGs in the atmosphere, and would only be motivated by a revolution in public and political climate risk perceptions.

- *Scenario 5: Technology induced transition to a low carbon economy.* This scenario is premised on big changes in the rate of development and or deployment of low-carbon technologies leading to a stabilization or reduction in atmospheric GHG concentrations, even in the absence of material carbon pricing.

Companies evaluating climate risks may wish to assign relative probabilities to the five scenarios introduced above as part of their risk management strategies. Such strategies have to accommodate potentially rapid future transitions from one scenario to another. Could the 'stay the policy course' scenario suddenly switch to the 'policy-induced atmospheric stabilization' scenario, or the 'return to 350 ppm' scenario in response to climate change itself? Are there circumstances in which the 'technology-induced transition to a low-carbon economy' scenario becomes more or less probable?

The business community has no prudent choice but to consider climate change as an integral part of corporate planning. For some companies, climate change and climate policy outcomes will create business opportunities. For the others, climate risk management strategies can already reduce companies' exposure and vulnerability to both climate change and climate policy. Climate risk positioning strategies can make companies 'response-ready' for climate risks that will evolve, or which cannot be materially or cost-effectively mitigated today. Companies with effective positioning strategies will be able to move more quickly than competitors when uncertainties around key variables are narrowed and thus enhance their competitive advantage.

The potential for business risk and business opportunity based on climate change and climate change policy is greater now than in the past, and will continue to grow as the gap between climate science and climate policy continues to expand. It is important that corporate risk management strategies keep up.

....

CHAPTER 2

Introduction

SCIENTISTS HAVE CALLED for a near-term reduction in global emissions of carbon dioxide (CO_2) of more than 70% to stabilize the concentration of CO_2 in the atmosphere. Meanwhile global CO_2 emissions, as well as emissions of the other so-called greenhouse gases (GHGs), continue to increase. While a political consensus exists for the view that exceeding 2°C of global temperature change would constitute 'dangerous anthropogenic interference with the climate system'[3] (the avoidance of which global governments are committed to through the United Nations Framework Convention on Climate Change), that amount of warming is already almost inevitable. More importantly, there is no global action plan in place to prevent much more dramatic temperature rises in coming decades.

Even as climate science has solidified, companies have been hearing for years that they don't need to know much about climate change science, they just need to recognize that 'the climate policy train is leaving the station, and you want to be on it'. This 'policy paradigm' of climate risk assumes that policy and regulation are the primary contributors to corporate climate risk, rather than climate change itself, and encourages policy-oriented risk responses. Correspondingly, the primary focus of corporate risk management activities has been to be at the policy table (rather than 'on the menu'), to measure and commit to reducing corporate carbon footprints, to anticipate the timing and magnitude of a future price on carbon, and to use carbon offsets to voluntarily reduce corporate or product-based emissions. Hundreds of corporate footprint reduction commitments and a slew of 'carbon-neutral' products and services have sprung up as a result.

Some 25 years after initial calls for broad-based GHG emissions reductions,[4] agreement on climate change policy to accomplish these reductions has proven an almost impossible nut to crack through domestic legislation or international negotiations. It's not for a want of trying; numerous policies intended to help reduce GHG emissions, and reduce or adapt to climate change are in place or being developed around the world. The problem is that these measures are unlikely to do more than scratch the surface of what scientists have said is necessary in order to materially reduce climate risk.

With the failure of national climate change legislation in the US, and the anticipated failure of international efforts to extend a meaningful version of the Kyoto Protocol, many companies are asking themselves: Climate risk? What climate risk? Companies should question, however, whether the 'policy paradigm' that underlies this conclusion, and that has guided corporate thinking for more than a decade, is actually the right risk management paradigm.

For example, does the growing disconnect between societal climate change risk and climate change policy have risk implications for business? How material to business is climate change itself, including all of the associated supply chain and brand risks? Is it reasonable to assume that if climate change makes itself increasingly felt it will become politically harder and harder to ignore, and that the risk of sudden and draconian policy risk will escalate? A gradual glide path to lower GHG emissions and toward a higher price on GHG emissions – long an objective of corporate efforts to influence climate policy – could be rendered moot if the public and policy-makers conclude that we have run out of time for gradual measures.

....

CHAPTER 3

The Elements of Climate Risk

THE BURNING OF FOSSIL FUELS was identified as a potential cause of future climate change more than 100 years ago.[5] At that time, however, no one foresaw the exponential rise in fossil fuel consumption that has accompanied global industrialization, or that the composition of the Earth's atmosphere would change at a rate the planet has never before experienced.

Why do scientists conclude that the climate is changing in response to human activities, and that such change poses risks? Risk is not about being able to predict the future with certainty; rather, risk is about understanding how exposed and/or vulnerable a system is to change:

RISK = EXPOSURE x VULNERABILITY

Perceptions of climate risk are therefore based both on 'knowns' and 'unknowns' regarding our exposure and vulnerability to climate change outcomes.

Climate change: What do we know for sure?

- We know that Earth supports life because trace levels of greenhouse gases (GHGs) (measured in parts per million), naturally warm the atmosphere by approximately 15°C through the so-called 'greenhouse effect'.[6] Without the natural greenhouse effect, Earth could not support life.

- We know that global temperatures have varied substantially over tens of millions of years of Earth history. But as illustrated in Figure 1, we also know that the last 10,000 years, during which human societies have evolved, have been remarkably climatically stable, with the global average temperature staying within a band of about 1°C.

- We know that human activities, primarily the combustion of fossil fuels, are releasing large volumes of CO_2 to the atmosphere, where it accumulates and contributes to radiative forcing[7] beyond that associated with the natural greenhouse effect.

- We know that GHG concentrations in the atmosphere are rising substantially. Concentrations of CO_2 have grown from 278 ppm (parts per million) at the onset of the Industrial Revolution to almost 400 ppm today;[8] indeed, concentrations of all six major GHGs have increased significantly.[9]

- We know that the last time global temperatures were 2°C and 3°C higher than 1900 levels, global sea levels were 4–6 meters and 20–30 meters higher than they are today, respectively.

- We know that conventional fossil fuels, if fully exploited, would increase atmospheric concentrations of CO_2 to more than 1000 ppm. Unconventional fossil fuels (e.g. oil shale and shale gas) have the potential to release even more CO_2 than conventional fossil fuels, as could (in CO_2-equivalent terms) the release of methane from melting permafrost or the release of huge quantities of methane stored as clathrates on the ocean floor.

....

FIGURE 1. Average global temperature over last 20,000 years

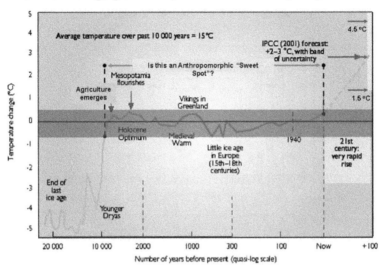

SOURCE: Adapted from *"Climate change and human health — risks and responses,"* published by WO in collaboration with UNEP and WMO 2003 and more recent data from IPCC 2007.

....

- We know that we're already seeing changes in local and global biological and other systems that are consistent with expectations of a higher-CO_2 and a warmer world, including reductions in the pH (acidification) of the world's oceans.

Climate change: What don't we know for sure?

- We don't know how quickly GHGs will accumulate in the atmosphere. It depends on many variables including population growth, economic activity, energy prices, and public policies.

Thus, we don't know whether CO_2 concentrations will rise to 450, 550, 650, or even higher ppm levels during this century.

- We don't know whether the oceans and other biological systems will continue to absorb, as they have been, approximately 50% of the CO_2 annually emitted to the atmosphere by human activities,[10] or whether a slow-down in ocean absorption will accelerate the growth in atmospheric concentrations.

- We don't know how sensitive the atmosphere is to changing levels of GHGs, e.g. through impacts on global cloud cover. Most global climate forecasts are based on assuming that a doubling of CO_2 concentrations will translate into approximately 3°C of average global temperature rise.[11] Scientists' estimates of climate sensitivity, however, range as high as 6°C.

- We don't know exactly how or when large-scale climate feedbacks (e.g. melting permafrost, Amazon forest dieback, or rapid melting of the Greenland ice pack) might be 'tripped'.[12]

- We don't know exactly how anthropogenic climate change is already manifesting itself, as compared to witnessing natural climatic variability.

- We don't know for sure why certain climate change impacts seem to be occurring more rapidly than forecast even just a few years ago (e.g. the rate of Arctic ice melt).[13]

These lists of what we do and don't know for sure reflect just a small slice of the available thinking around climate change. What is notable from even this short list is the range of potentially risky outcomes associated with these uncertainties:

- What if GHG concentrations in the atmosphere start growing more rapidly?

- What if the sensitivity of the atmosphere to changing GHG concentrations is higher than assumed in the global models being used in political decision-making?

- What if suspicions prove true that climate change is already stressing global food production, leading to declines in reserves and significant price increases?

- What if climate 'tipping points' lead to step changes in observed climate change?

Rather than generating a policy debate over how much climate change risk is acceptable, however, the prevailing public debate has been characterized by arguments over *exactly* what is happening today, whether it is *definitely* attributable to climate change or might reflect natural variability, and *exactly* what will happen in the future as a result of anthropogenic climate change. This political discourse runs contrary to the whole notion of risk assessment and management under conditions of uncertainty.[14]

The fact that sea levels were 4–6 meters higher than today the last time global temperatures were 2°C higher than today, and that fact that we are expected to witness that magnitude of temperature change in the next few decades, doesn't necessarily mean we should anticipate a near-term catastrophic rise in sea levels. Nor do many scientists share scientist Jim Hansen's belief that current trends could lead us to trip the ultimate climate change tipping point – the Venus effect – referring to the runaway accumulation of CO_2 in the Earth's atmosphere.[15] But such observations and hypotheses by credible observers do suggest that we should be on the watch for risky surprises as climate change continues to manifest. As shown in Figure 1, human societies have evolved within a narrow 1°C temperature band, a band we are now beginning to move outside of in an aggressive way. As we stray farther outside this band we may find ourselves much more susceptible to surprises than we have yet anticipated. Figure 2 illustrates that the level of scientific concern over degree to degree changes in global temperature, including with respect to the risk of 'major discontinuities', increased significantly between the Third Assessment Report of the Intergovernmental Panel on Climate Change (IPCC) in 2001 and the IPCC's the Fourth Assessment Report in 2007.[16] (See Figure 2)

In considering climate risk, we also have to remember that climate change will not be noticeable to us through changes in global averages, whether in temperature, precipitation, or sea level rise. Only global satellites will witness global climate change. Climate change and its attendant risks will manifest themselves primarily at the local and regional levels, and in ways that may diverge substantially from 'the average':

- ✓ Localized temperature changes will often significantly exceed average global temperature changes, with localized energy, health, and economic implications.

- ✓ Changes in water availability and timing will vary substantially, threatening local crops, drinking water supplies, and energy production.

- ✓ Sea level rise will in many areas cause permanent losses of wetlands and shoreline, worsened storm surges, and saltwater intrusion into freshwater aquifers.

FIGURE 2. Changing perceptions of risk across five degrees of change

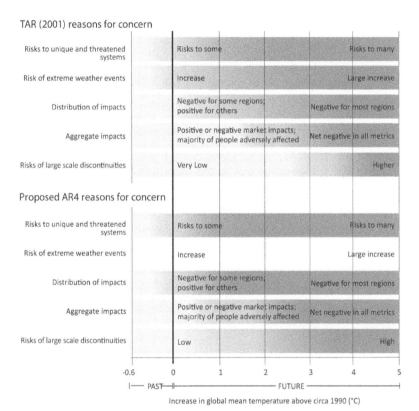

TAR (2001) reasons for concern

Risks to unique and threatened systems	Risks to some			Risks to many
Risk of extreme weather events	Increase			Large increase
Distribution of impacts	Negative for some regions; positive for others		Negative for most regions	
Aggregate impacts	Positive or negative market impacts; majority of people adversely affected		Net negative in all metrics	
Risks of large scale discontinuities	Very Low			Higher

Proposed AR4 reasons for concern

Risks to unique and threatened systems	Risks to some			Risks to many
Risk of extreme weather events	Increase			Large increase
Distribution of impacts	Negative for some regions; positive for others		Negative for most regions	
Aggregate impacts	Positive or negative market impacts; majority of people adversely affected		Net negative in all metrics	
Risks of large scale discontinuities	Low			High

-0.6   0   1   2   3   4   5

|—— PAST —|—————————— FUTURE ——————————|

Increase in global mean temperature above circa 1990 (°C)

SOURCE: Mabey, N., et al. 2011. *Degrees of Risk: Defining a Risk Management Framework for Climate Security* (London: Third Generation Environmentalism Ltd).

✓ Ocean acidification will interfere with shell formation and crustacean reproduction, with localized impacts on fisheries and livelihoods, while also threatening the world's coral reefs.

✓ More frequent and wide-ranging localized crop failures, more severe droughts, and rising food prices could lead to greater national and regional instabilities around the globe, and increase the potential for international conflict.

An obvious question for business observers is just how sure scientists are about climate change and the existence of physical climate risks. There is no question that consensus exists among the scientific community that anthropogenic climate change poses serious risks, notwithstanding a range of continuing uncertainties regarding the magnitude, pace, and impacts of climate change. In interpreting the continuing scientific debates over these aspects of climate change, we have to remember that there is virtually nothing that scientists agree upon universally. This should not be interpreted as somehow undermining scientific certainty about climate change; there's almost nothing that scientists consider 'certain' in the way the term is commonly used.

That said, few scientists would dispute the possibility that climate change will prove less severe than currently feared. Current science, however, suggests that the probability of such an outcome is quite low; considerably lower in fact than the probability that today's prevailing climate change forecasts significantly *understate* climate risks. As shown in Figure 3, the 'long tail' of the climate risk distribution has expanded based on improving knowledge as well as the 'what we don't know for sure' variables introduced above. Perhaps because most scientists are loathe to be perceived as alarmists, however, there has been relatively little public discussion of the probability of 'long tail' climate change risks. (see Figure 3)

FIGURE 3. The long tail of scientific climate uncertainty.

Based on recent observations compared to climate model projections, the probability distribution of climate change outcomes appears to be biased systematically toward more severe outcomes.

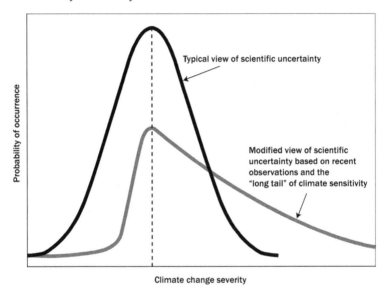

SOURCE: Adapted from Gulledge, J. 2008. Climate change risks in the context of scientific uncertainty. In K.M. Campbell and J. Price (eds) *The Global Politics of Energy*, pp.115–132. Queenstown, MD: The Aspen Institute

Societal vs Business Perspectives on Risk

It can be tempting to characterize 'risk as risk', and to argue that we should all perceive and respond to risk similarly. Risk, however, is always to some extent in the eye of each beholder; each of us will process risk-relevant information differently based on our understanding of our exposure and vulnerability to risk factors, as well as on our experiences and backgrounds.

We should, however, differentiate broadly between societal and business perspectives on risk. Societally-relevant time scales when it comes to risk perception are usually much longer than corporate time-scales, where even a five-year outlook can seem like the distant future. Societal damage functions are also usually much broader in assessing the impacts and consequences of risk, and in the case of climate change extend to such impacts as world food supply disruptions, acidified oceans, migration of climate refugees, and the potential for heightened international tensions and conflicts. Societal risk management processes are supposed to respond to these risks through societal action, e.g. through policy and regulation.

Corporate risk assessment, on the other hand, tends to focus on near-term operational impacts and the 'needs' of shareholders and other stakeholders. Business-relevant time-scales are shorter than societal time-scales, and businesses tend to discount future costs and benefits at a higher rate than in societal risk analysis. The implications of this should not be ignored; at a 20% discount rate, problems like climate change can literally seem to 'disappear' as even extreme future costs are 'discounted away'. Businesses also have little ability to fix economic externalities that drive problems like climate change. The bottom line is that corporate risk managers tend not to focus on uncertain and future problems over which they have little control, and that are seen as beyond their planning horizons.

As a result, it is unrealistic to expect business to look at climate risk through a societal lens, much less should business action on climate change be expected to substitute for societal risk management. The business sector can play a huge role in helping advance climate change mitigation and adaptation objectives, but it has to be within a societal risk management context. Companies cannot individually internalize economic externalities and hope to remain competitive against companies that don't. Nor can companies voluntarily implement societal discount rates in their decision-making, and hope that investors will reward them.

Some observers would argue that the business community has a greater obligation to help address climate change than might appear to be suggested here. We are not weighing in here on debates over corporate responsibility and intergenerational ethics. We are simply noting that it is not realistic to suggest that the private sector should or can substitute for the world's governments in managing societal climate risks. This is highly relevant for how the business community perceives and responds to climate risks. If the business community can't directly address climate change, and society doesn't, then the business community has to anticipate greater levels of climate risk.

....

CHAPTER 4

The Challenge of Societal Climate Risk Management

THE SECTION ABOVE DESCRIBED the problem of climate change as a societal risk, and explored why societal and business perceptions of climate risk can differ fundamentally. Societal risk management has been reasonably successful for many environmental risks, from air pollution and oil spills to the ozone hole. Why, then, after 25 years of active engagement, has the problem of climate change not been tackled by policies or other measures that can reasonably be expected to have a material impact on the problem?

Some of the reasons for our difficulties in this area are quite obvious, others more subtle. What is safe to say is that the problem of climate change is not characterized by any of the attributes that make a problem relatively easy to solve. Quite the contrary, as explored below:

- Almost all human activities, from energy consumption to food production, contribute to GHG emissions. Compare this to efforts to address the ozone hole, where just a few chemicals being produced by a small number of factories in a small number of countries had to be tackled.

- There are no silver bullet technological solutions for GHG emissions. Scrubbing sulfur dioxide from the flue gas of coal-fired power plants, for example, radically reduces sulfur emissions at modest cost. The same is not true of CO_2.

- Climate change occurs over very long periods of time and is almost invisible to human perception against the backdrop of natural variability. Compare this to virtually all environmental problems that we've successfully tackled, from oil spills to severe air pollution.

- Perceptions of climate change risk suffer at the hands of typically employed risk assessment tools and metrics such as cost–benefit analysis. Even the costs of extreme climate change, occurring more than 10–20 years in the future, can appear immaterial if discounted at sufficiently high discount rates.

- The costs of mitigating climate change tend to be front-loaded, and could be very expensive if they are undertaken suddenly. Large quantities of capital stock could be 'stranded' if not enough time is provided for technologies and systems to evolve that would cushion costs.

- Climate change is a difficult concept to communicate. Our primary scientific measure of climate change – tenths of degree changes in the world's average global temperature – is unsuited to the realities of human risk perception and hides projected localized effects.

- The magnitude of what scientists believe is needed to address climate change is immense. Simply stabilizing the concentration of carbon dioxide in the atmosphere would require greater than 70% reductions in global CO_2 emissions. Yet the Kyoto

Protocol's targets – while widely perceived as having been costly and challenging – only called for very modest reductions from industrialized countries against a 1990 baseline, while global emissions actually continued to increase.

- Climate change has all the attributes of a 'tragedy of the commons'. The benefits of mitigating climate risks are widespread, but the impacts of aggressive climate policies would fall heavily on certain countries and certain sectors, including those with political veto power. The countries perhaps most immediately facing global climate change, including small island states threatened with being wiped out by sea level rise, don't actually emit many GHGs.

These are a few of the variables that make climate change risk management so difficult. When considering the perceived near-term costs of tackling climate change, against the hoped-for long-term benefits, it should be no surprise that material climate change policy has proven so difficult when political decision–makers have electoral outlooks of two to six years.

A few particularly important barriers to climate action are profiled in somewhat more detail below, given their importance to understanding the probability distribution of business climate risks.

Climate Change Cost–Benefit Analysis

Conventional economic thinking poses a challenge to the perceived need for climate change risk management. Put most simply, the costs of aggressively mitigating climate change are biased toward the near term, while the benefits of aggressively mitigating climate change are biased toward the long term. Taking into account the time value of money, climate change can quickly seem to become an even more invisible problem than it already is.

In 2006, the British government published the so-called Stern Review (commonly named for the head economist, Nicholas Stern, in charge of the review group putting together the report).[17] The Stern Review concluded that climate change threatens to reduce future global GDP by 5-20% in perpetuity, and that we should be willing to pay a great deal of money today to avoid such an outcome. The Stern report was criticized at the time for basing its findings on the use of a very low discount rate. Employing typical societal discount rates of 3-4% (much less typical private sector discount rates of 10-20%), significantly changes the Stern Review results.

But this begs the question, according to some economists.[18] For them, the key question is whether typical cost-benefit analysis is appropriate for a problem with the characteristics of climate change. We usually don't intentionally and explicitly make decisions that would dramatically and negatively affect future generations just because the net-present-value (NPV) of such decisions looks favorable to today's decision-makers. Nuclear waste disposal, generally evaluated against the requirement that the waste be sequestered for thousands of years, is an interesting juxtaposition to climate change policy discussions.

There is another side to the economic debate over climate change. Many economists argue that focusing on total discounted dollars is the wrong way to think about climate risk. Even modest reductions in future GDP caused by climate change mitigation efforts can add up to a large figure in absolute dollars. But what if the question is phrased differently - i.e. is it worth accepting a 95% increase in 2030 GDP, as opposed to a 100%

increase in 2030 GDP, in order to make the investments necessary to manage climate change risk? This puts the question into an entirely different light allowing, it is argued, for a more risk- and values-based discussion. The Stern Review, for example, contrasts its estimate of 5–20% of loss in annual GDP to a 1% of annual GDP investment required to avoid negative climate change outcomes through effective societal risk management.

Other economists argue that there is a good chance the world economy would actually be better off in the future by tackling the climate change problem today, even without factoring in the costs of future climate change. They argue that moving to replace the high environmental, social, and security costs of today's global energy systems with alternative energy systems would pay off quickly.

Economic analysis of climate change risk at the local and project-based level is still relatively rare, primarily because the ability to forecast the localized impacts of climate change is still evolving. One credible economic analysis, looking at hurricane risk to New York City, concludes that by 2030 climate change will increase the risk of a category 4 or 5 hurricane hitting New York City by as much as 25–30%,[19] and that the city's vulnerability to such an event is so great that the city should be willing to pay more than $47 billion per year in 2030 if it were possible to avoid the increased risk. That is more than 50% of New York City's annual expenditure budget ($60 billion in 2009),[20] just to purchase a climate change insurance policy against climate change-induced increase in hurricane risk.

The same analysis concluded that investors funding hurricane-susceptible infrastructure should be prepared for a significant decline in their expected rates of return from such investments, for the same basic reason of increasing extreme event risk. For infrastructure with an estimated 12% return based on historical climate, for example, the analysis suggested that expected future rates of return may prove closer to 4%.

Mitigation vs. Adaptation – Is That the Question?

There is a growing controversy over the path of climate change risk management. Risk management historically has been conceived of largely as referring to the mitigation of climate change through reduction of GHG emissions sources and enhancement of sinks. Mitigation has proven a tough nut to crack, however, and many observers are beginning to characterize adaptation to climate change as a more pragmatic risk management strategy.

Clearly, adaptation will be an important component of any response to climate change.[21] As suggested by the recent report *Degrees of Risk*, 2–3°C of global temperature change is now the 'best case', whereas the worst case is 6–8°C.[22] As shown in Figure 4, these ranges are based on alternative assumptions regarding climate sensitivity and the success of climate change mitigation policies. Even the 'best case' of 2–3°C will require significant adaptation measures. (See Figure 4)

Adaptation, by limiting the consequences of climate change, certainly qualifies as a climate risk management strategy. But the key question in characterizing adaptation as a potentially superior risk management strategy is the ultimate outcome. The potential effectiveness of adaptation efforts, at least in terms of significantly moderating the consequences of climate change for the bulk of the world's population,

....

FIGURE 4. Four-cell climate change scenarios to 2100.

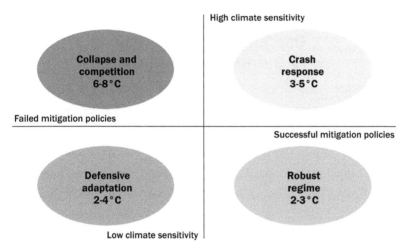

SOURCE: Mabey, N., et al. 2011. *Degrees of Risk: Defining a Risk Management Framework for Climate Security* (London: Third Generation Environmentalism Ltd).

....

becomes less viable as the magnitude of climate change increases. As was suggested in Figure 2, unmitigated climate change would disrupt the ecosystems and infrastructure on which billions of people depend. The world's resources and ecosystems are already far more stressed today than in the past, and there is far less potential for human populations to move in response to a shifting climate than has been the case in the past. The prospect of huge numbers of climate refugees is a daunting one and the potential for global conflicts is large.[23]

As such, a pure adaptation approach to climate risk management is plausible for just a small percentage of the Earth's population. Moreover, adaptation does not address many secondary risks of climate change to food supplies and ecological systems. From a risk perspective, adaptation should not be treated as 'the' answer to climate risk management.

Perceiving Climate Risk – How Our Brains Get in the Way

Climate risk has been a source of active discussion for 25 years; yet clear policy to address the climate change problem has yet to appear. It is easy to attribute this failure to the 'tragedy of the commons', the existence of economic externalities, or the political influence of fossil fuel interests. But research over the last 20 years has also opened a growing window of investigation into the role of human decision-making, in how we perceive and respond to different kinds of risks.[24] It is increasingly clear that the human brain can operate to the detriment of logical and rational thinking when it comes to complicated problem-solving.

One explanation of this is that human society, and the nature of the problems we are trying to address, has evolved more quickly than the physical capacity of our brains to understand and respond to risk. As Daniel Kahneman put it in his 2011 book *Thinking, Fast and Slow,* humans suffer from the 'what you see is all there is' (WYSIATI) phenomenon. WYSIATI theory suggests that human decision-making is based primarily on *Known Knowns*, namely phenomena we have already observed, and rarely considers *Known Unknowns*. Most importantly, human decision-making appears almost oblivious to the possibility of *Unknown Unknowns*, including the kinds of climate change tipping points discussed above. Brief profiles of specific cognitive biases are introduced below, with short discussions as to how they are potentially relevant to perceptions of climate risk.

- The availability heuristic (http://en.wikipedia.org/wiki/Availability _heuristic) is a way of describing how humans make judgments. We think we're rational creatures, using all available information to make decision. In fact, we make judgments based on what we best remember, which tends to be very recent experiences. Thus, in the days or weeks after a major earthquake homeowners will flock to buy earthquake insurance. Over time, the number of people buying earthquake insurance declines, even as the objective likelihood of an earthquake increases as tectonic stresses increase. The *availability heuristic* has major implications for how we think about a risk like climate change that to date is difficult to differentiate from natural climate variability.

- The optimism bias (http://en.wikipedia.org/wiki/Optimism_bias) is used to describe the human tendency to expect things to turn out better than data-supported forecasting may suggest. Regardless of the data presented, we tend to believe that the future will be much better than the past. The bias holds true across ethnic and socioeconomic groups. Taken collectively, people can grow pessimistic about world economies or the future of their country, but individually, they will tend to believe their lives will get better.[25] This bias also interferes with evidence-based risk perception for a problem like climate change.

- The *neglect of probability* bias (http://en.wikipedia.org/wiki/ Neglect_of_probability) is another bias that kicks in under conditions of uncertainty, based on the difficulty our brains have in working with probabilities. In a field like climate change, as data intensive as it is, the 'neglect of probability bias' suggests that much more focus needs to be made on communicating climate risk in a way that avoids the potential for this bias to interfere.

- *'Patternicity'* is another cognitive bias with huge implications for climate risk perception.[26] The search for patterns in everything around us has clear evolutionary benefits, but it is exactly that kind of pattern-seeking that has helped political discourse get bogged down in debates of whether climate change is or isn't already happening.

This short review just scratches the surface of the risk perception and risk management implications of cognitive biases as suggested in recent research, but proposes that much more policy and communications attention will be needed if these barriers are to be overcome when addressing climate risk perception and management.

....

CHAPTER 5

Scenarios of Climate Change Risk

ONE COULD EASILY CONCLUDE that the climate change problem is basically impossible to solve and that risk management is largely a function of waiting for climate change to happen. But predictions of the future are notoriously unreliable.[27] It is particularly difficult to reliably forecast specific climate risk outcomes given the sheer number of and uncertainties associated with physical, economic, and political variables.

The entertainment field and mass media have produced multiple movies and books that have painted wide-ranging scenarios regarding climate change and climate risk. Many scenarios painted in these venues are pessimistic (e.g. the 2004 film *The Day After Tomorrow*) and moralistic (e.g. how could today's generation have let this happen?). But others are much more optimistic, envisioning a relatively painless transition to a 'low carbon future' through means and technologies already at our disposal.

The 'opportunity storyline' is attracting a large following among business and environmental groups seeking to motivate corporate action on climate change. Proponents of this opportunity-based approach believe that it will motivate corporate action as well as reduce business opposition to material climate change policy. A casual observer of today's corporate climate change strategy literature might think that global climate change is a good thing, given all the 'opportunities' it is creating.

Risk and opportunity are two sides of the same coin. In a disaster-prone scenario, disaster-response companies can anticipate big profits. Any major effort at climate risk management, while proving fundamentally disruptive to certain business sectors, will create opportunities for other sectors. This should not be mistaken, however, for the ability of companies to find win-win solutions across the board, nor for their ability to choose whether they will ultimately be winners or losers. The reality is that individual companies may be able to influence the degree to which they win or lose under alternative climate risk and climate policy scenarios through risk management and risk-positioning efforts today, but relatively few can determine whether they are on the winning or losing side of the ledger (without fundamentally changing their business model).

From a risk perspective, it is easy to overplay the business opportunity side of climate change policy and risk management for today's dominant industries. With scientists calling for dramatic measures to reduce dangerous climate change, the gap between science and policy is enormous and continues to grow.

This almost certainly contributes to future business risk, as further explained below. Most companies today are in all likelihood substantially underestimating the range of potential business climate risks. This is not to argue that companies are ignoring climate change; more and more companies are explicitly seeking to address climate change through a variety of corporate programs, most prominently through sustainability and corporate social responsibility strategies. But how many companies would be prepared for climate change or climate change policy outcomes under several of the scenarios introduced below, and the attendant impacts on their operations and supply chains?

From a business risk standpoint, it is useful to characterize a range of climate change scenarios against which potential business risks can be compared and evaluated. Five scenarios are profiled below, representing a large part of the potential distribution of climate change and climate policy outcomes:

Scenario 1: Issue collapse. The prospects of climate change, and the pressure for policy action on climate change, could come to an end. This would require a radical reversal in the current state of climate change science. Based on current knowledge, this appears a particularly low-probability scenario.

Scenario 2: Stay the policy course. Under this scenario, numerous policies and measures continue to be pursued domestically and internationally, accompanied by relatively modest policy and regulatory commitments. This scenario can be thought of as equivalent to an explicit or implicit carbon price of US$5–30/ton of CO_2, which would not be sufficient to stabilize GHG concentrations in the atmosphere. As a result GHG concentrations continue to rise in this scenario, albeit at a modestly lower rate that under Scenario 1. Assuming no fundamental changes in public perceptions of climate change, the probability of this scenario in the near term is relatively high.

Scenario 3: Policy induced atmospheric stabilization of CO_2. This scenario is based on political will coalescing to pursue the aggressive emissions reductions and technology development initiatives that would be necessary to stabilize GHG concentrations in the atmosphere. This scenario can be thought of as equivalent to an explicit or implicit carbon price of US$50–150/ton of CO_2, which would have transformative implications for major sectors of the global economy. GHG emissions fall dramatically in this scenario, and atmospheric concentrations of CO_2 stabilize within the 450–650 ppm range. This scenario would require a fundamental change in public and political perceptions, perhaps brought about climate change itself, and adaptation is an important part of the risk response picture. The probability of this scenario appears relatively low in the near term, but probably grows significantly in the mid to long term.

Scenario 4: Policy induced return to 350 ppm CO_2. This scenario extends Scenario 3 by envisioning a return to an atmospheric concentration of CO_2 of 350 ppm, almost 50 ppm below today's levels. Some observers are advocating this scenario due to concerns that today's CO_2 concentrations may already be too high to protect the world's oceans and other ecological systems. This scenario would only result from a revolution in public and political climate risk perceptions.

Scenario 5: Technology induced transition to a low carbon economy. This scenario arrives at roughly the same end-point as Scenarios 3 or 4, but without having to assume the disruptive climate change and associated climate policy interventions anticipated in those two scenarios. This scenario is premised on big changes in the rate of development and or deployment of low-carbon technologies, even in the absence of material carbon pricing. As long as GHG emissions represent an economic externality and continue to be largely un-priced, few climate change or technology observers see this as more than a very low probability outcome.

Climate Scenario Wildcards

Companies evaluating climate risks may wish to assign relative probabilities to the five scenarios introduced above as part of their risk management strategies. Such strategies have to accommodate potentially rapid future transitions from one scenario to another. Could the 'stay the policy course' scenario suddenly switch to the 'policy-induced atmospheric stabilization' scenario, or even the 'return to 350 ppm' scenario? Are there circumstances in which the 'technology-induced transition to a low-carbon economy' becomes much more probable? Two key wildcards are particularly relevant in thinking about these questions.

i. Climate black swans

Technically, black swan events represent outcomes that simply weren't known to be possible, or were expected to be so rare that they cannot reasonably be predicted.[28] For purposes of this climate risk discussion, we expand the scope of black swan events to encompass events that we may recognize as possible, but which are generally not considered in the development of business risk management strategies. The distinction between these two types of outcomes is illustrated in Figure 5.

Potential black swans when it comes to climate risks include certain policy outcomes, e.g. the imposition of dramatic emissions reduction mandates, or much higher carbon prices than generally anticipated. They include certain technology outcomes, e.g. dramatic breakthroughs in new technologies or in the costs of existing technologies. They include climate change outcomes themselves, such as potential tipping points like large-scale methane releases from permafrost, the shifting of ocean

....

FIGURE 5. Typical vs fat-tail climate risk planning.

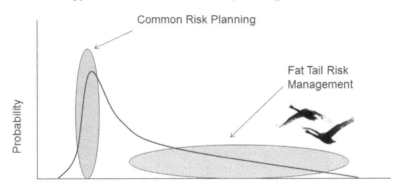

SOURCE: The authors.

....

currents, or the rapid loss of the Greenland ice sheet. A range of such tipping points is shown in Figure 6.

Black swan events such as these could have fundamental implications for the likelihood of the five risk scenarios introduced above. The actual impact of any given black swan event or combination of events, however, will be determined by a more complicated variable, the Climate Response Tipping Point (CRTP).

ii. The Climate Response Tipping Point

The Climate Response Tipping Point (CRTP) acknowledges the complicated and contrasting pressures advocating for and against a risk-based societal response to climate change. The CRTP is premised on the idea that at any given time there is the potential for a climate

FIGURE 6. Large-scale potential climate tipping points.

Map of potential policy-relevant large-scale tipping elements in the climate system overlain on global population density.

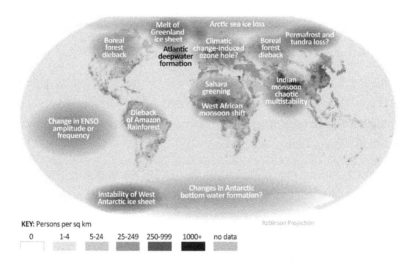

KEY: Persons per sq km

| 0 | 1-4 | 5-24 | 25-249 | 250-999 | 1000+ | no data |

SOURCE: Lenton, T.M., Held, H., Kriegler, E., Hall, J.W., Lucht, W., Rahmstorf, S. and Schellnhuber, H.J. 2007. Tipping elements in the Earth's climate system. *Proceedings of the National Academy of Sciences USA* (Volume 105, Issue 6): 1786–1793. © Copyright (2007) National Academy of Sciences, USA.

change event or combination of events to 'tip' public and policy-makers' perceptions of climate risk sufficiently to overcome the barriers that have to date impeded climate policy development.

The CRTP does not represent a fixed point in the future. We do not know for sure what climatic event(s) would be sufficient to trip the CRTP today, and indeed the constellation of events that would trip the CRTP is always fluctuating. Variables contributing to the nature of the CRTP at any given point in time include political, policy, economic, technology, and social variables. Significant technological advances, for example, would make it easier to trip the CRTP because climate change action would be perceived as less costly than it is today. Global economic disruptions, however, tend to make it harder to trip the CRTP because the focus on near-term costs and benefits intensifies.

Because the CRTP is not a fixed quantity, it can be intentionally shifted. If the CRTP can be triggered sooner by less severe climate events, e.g. through improved risk literacy and climate risk communications, societal costs of climate change will be reduced.

From a business risk perspective, the CRTP illustrates the 'double-whammy' nature of business climate risk under Scenarios 3 and 4 above. If the CRTP is tripped, it suggests that businesses will be grappling with the direct effects of climate change in operations and on supply chains, and will be investing resources in adapting to those changes. But if the CRTP is tripped, it also suggests that more climate policies will quickly follow, potentially creating a more draconian policy environment than businesses currently expect.

One can reasonably disagree as to the likelihood of the CRTP being tripped in the foreseeable future, either because we won't witness the necessary climatic events or because the CRTP proves too high a hurdle. Today's corporate climate risk strategies implicitly treat the CRTP as a black swan, either assuming that the CRTP doesn't exist or that it won't manifest itself during timeframes relevant to business planning. Yet current climate science projects that dramatic climate change events are virtually inevitable. As a result, draconian policy responses are almost inevitable as well. There is considerable uncertainty about the timing of these outcomes, but it is hard to argue based on current science that these outcomes are not possible within timeframes relevant to business planning, particularly with respect to long-term infrastructure and other investments. Even a modest probability of the CRTP being tripped during the business planning horizon of significant sectors or companies constitutes a business risk that deserves corporate risk management attention.

....

CHAPTER 6

Bounding Business Climate Change Risks

CLIMATE CHANGE HAS BEEN RECOGNIZED as a business risk, at least for major greenhouse gas (GHG) emitters, for the last 20 years.[29] Today, climate change and related risk variables, e.g. water scarcity and extreme events, increasingly rank toward the top of the business risk list published annually by the World Economic Forum (see Figure 7).[30]

It is not difficult to identify how climate change (and climate change policy) can translate into business climate risk:

- *Physical risk*, including direct impacts of climate change on a company's operations, supply chains, financial performance, and employee health, among other variables. It will be difficult for companies on their own to materially lessen the physical risks of climate change, but they can reduce their vulnerability to such changes through adaptive measures internally and through their supply chains.

- *Brand risk*, including the impact on corporate competitiveness of consumer and stakeholder perceptions of the company. Brand risk (and opportunity) has been a primary driver to date of corporate responses to climate change. But such responses can create more risk than they alleviate, e.g. if it leads to 'greenwashing' claims.

FIGURE 7. World Economic Forum 2011 global risk landscape.

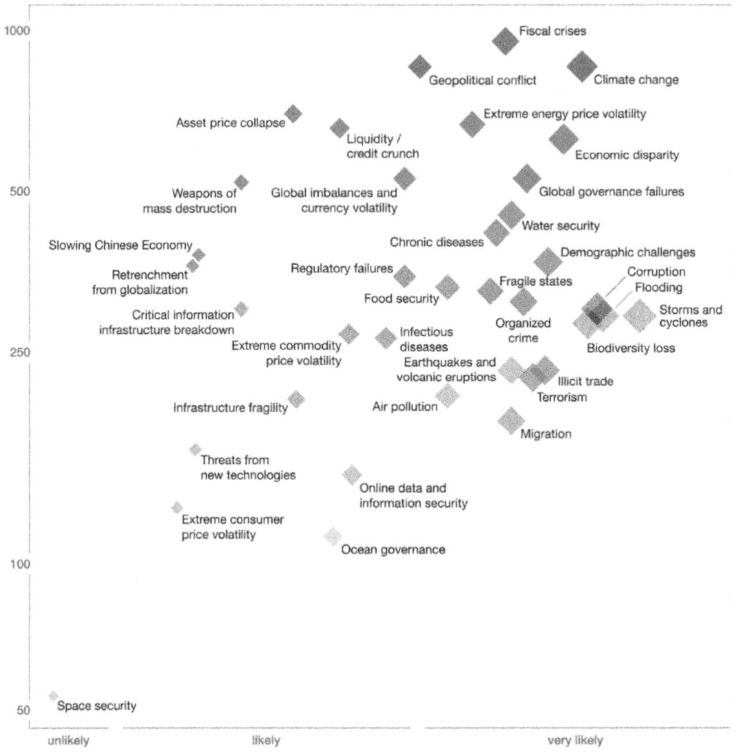

SOURCE: Global Risks, 2011. 6th Ed. World Economic Forum. Available at: http://reports.weforum.org/global-risks-2011/

Overall, climate brand risk can be mitigated through measures like corporate transparency and disclosure, corporate leadership in the pursuit of innovative climate risk best practices, and other steps that contribute to being perceived as a good climate citizen.

- *Policy risk*, including the impacts of climate change policy and regulatory mandates on a company's competitiveness, and the susceptibility of corporate business models to a governmentally-generated price on carbon. Many companies have responded to policy risk through policy engagement, and efforts to measure and reduce their GHG footprints. Some companies have chosen to address policy risk by working to impede climate policy. Most companies advocate near-term but gradual policies and measures that are supposed to provide investment guidance but avoid sudden business dislocations and stranded costs. The biggest source of policy risk would be tripping the CRTP.

- *Structural risk*, including the impacts of climate change-influenced market forces on the supply of and demand for a company's products and services, or investor perceptions of its likely long-term performance in a carbon-constrained world. This includes the susceptibility of corporate operations to international instability fueled by the implications of climate change, from food and water shortages to population migration. Structural risk also includes susceptibility of corporate business models to the rate and nature of technology innovation, whether as a creator of business opportunity or a creator of competitive disadvantage.

- *Liability risk*, including litigation or legislation that could result in retroactive or future corporate liability for GHG emissions. This includes the susceptibility of corporate business models to charges of inadequate climate risk disclosures. This risk, while related to policy and regulatory risk, is listed separately as it could manifest itself through litigation, rather than through a policy process. The types of risk management measures that will prove effective in managing brand risk could help manage liability risk as well.

The relative importance of these risks will differ not only based on which future climate scenario one envisions, but also according to the business sectors and specific company one is analyzing. See Table 1 for an assessment of the climate risk facing a range of business sectors. Note that most investment analysts producing such assessments focus on policy risk rather than the physical risks of climate change.

TABLE 1. Industry winners and losers

Category	Short	Medium	Long
Electric-Generating Equipment	++	+++	+++
Energy-Efficiency Enhancing Technologies	++	+++	+++
Engineering & Construction	+++	+++/−	+++
CO2 Transport, Injection and Storage	+	++	+++
Electric Utilities	++/−	+++/−	++
Transportation	++/−	+++/—	+++/—
Commodity-Related	+/-	++/-	+++/−
Electricity-Intensive Industries	+/-	+/−	+/−

SOURCE: AllianceBernstein 2008. *Abating Climate Change: What Will Be Done and the Consequences for Investors,* p. 86 (New York: AllianceBerstein LP).

Additionally, future emissions reduction mandates will likely be pegged against companies' new and improved future emissions baselines, rather than their original emissions baselines. The well-known metaphor 'no good deed goes unpunished' reflects this aspect of business risk.

No good deed unpunished?

Companies have been pursuing voluntary climate action measures for 25 years. The first corporate carbon offsets were undertaken by AES Corp., a global energy producer, in 1989. AES invested in offset projects in Latin America, which were intended to offset new power plant emissions in the US long before emissions reduction mandates were on the table.

Corporate footprint reduction efforts also date back to the early 1990s and electric utility commitments to the Climate Challenge Program of the US Department of Energy which began in 1994. The first carbon-neutral company commitment was in 1996; the first carbon-neutral products date from around the same period.

Most of these initiatives were undertaken to hedge one or more business climate risks, primarily policy risk; or to take advantage of branding and market-building opportunities through early actions. In some cases, both rationales may have applied.

The effectiveness of many of these initiatives can be questioned. In most cases it is not clear whether corporate initiatives to manage climate policy risk have actually contributed to that goal. In other cases, it has turned out to be far more difficult than companies had realized to sway consumer behaviors based on climate change motivations, and to translate changed behavior into corporate competitive advantage.

From a risk perspective, two potential outcomes of early corporate actions are particularly interesting:

- Are early corporate actions to reduce emissions or hedge future risk rewarded, ignored, or penalized? Do companies that take early actions (i.e. in advance of any regulatory requirement) to reduce their GHG emissions risk changing the emissions baseline used to calculate their allowed emissions under future regulatory mandates, thus actually increasing their regulatory risk?

- Do early corporate actions risk alienating climate-focused consumers more than they contribute to attracting new consumers? This has certainly been documented in some carbon offset cases, where well intentioned companies have been caught up in brand damage resulting from charges of 'greenwashing'.

Every year, more companies face shareholder resolutions or other public pressures to act on climate change. In responding to these pressures, companies walk a fine line. Determining what actions to take is not always easy. Companies want to appear responsive, but are worried that no good deed may go unpunished.

While these business risks may seem straightforward, most businesses make a series of implicit or explicit risk assumptions that tend to minimize perceived risk and bias corporate strategies towards 'wait and see' responses. Assumptions contributing to this outcome include:

- The assumption that the atmosphere's sensitivity to anthropogenic climate forcing will be at the lower risk end of the overall probability distribution, suggesting that climate change impacts will be slower to be felt, more gradually felt, and less cumulatively important than the 'long tail' of the climate risk probability distribution would suggest.

- The assumption that the risks of climate change itself will manifest themselves slowly and linearly, rather than through 'black swan' game-changing events. In other words, so-called climate change 'tipping points' will not manifest themselves during a time period relevant to business risk analysis.

- The assumption that the future of climate policy will look much like the past, meaning a lot of talk and little action, and that policy will be slower to develop, more gradually implemented, and less draconian than the 'long tail' of the climate risk probability distribution would suggest.

- The associated assumption that market structure, branding, and liability risk outcomes associated with climate change and climate change policy will also be in the lower-risk end of the probability distribution of potential outcomes.

- The assumption that climate risks are really no different from other business risks, notwithstanding the potential scale and irreversibility of the climate change problem.

- The assumption that it's OK to focus on 'most likely' risk outcomes, rather than on outcomes located in the more dangerous 'long tail' of the risk probability distribution.

- The assumption that it's OK to focus on the specific risks a company feels most able to address, rather than on exposure and vulnerability to the full range of climate risks.

Companies implicitly or explicitly employing one or more of these assumptions in their risk management strategies are rolling the climate risk dice and assuming that the dice are 'loaded' toward lower-risk outcomes. There are multiple reasons that companies want to make this assumption; these reasons are compounded by cognitive barriers in human decision-making. We have a deep-seated desire to know the future, a deep-seated bias towards optimism when doing so, and a deep-seated aversion to linking our personal or business activities to negative future outcomes.[31] When it comes to risk perception and management, this is a challenging combination of forces to counter through objective climate change risk assessment and management.

It is certainly possible that even fair dice being rolled just a few times can generate the outcomes assumed by business. But there is, of course, a much higher probability that they won't. Companies are thus taking more climate risks than they realize. This conclusion doesn't imply that those risks are always material, or that companies' futures depend on managing those risks. Whether the 'true risks' are actually material depends on many variables, including the sector within which the company operates and its exposure and vulnerability to the risks identified above. Only a materiality assessment can shed adequate insight into this question at a company-specific level.

....

CHAPTER 7

Assessing Corporate Climate Risk

AS JUST NOTED, THE EXISTENCE of potential risk does not automatically imply that a company needs to make managing that risk a priority. To conclude that a changed risk response makes sense, corporate decision-makers need to affirmatively answer two key questions:[32]

- Is the problem worth it?

- Can I succeed?

There is no one-size-fits-all approach to answering these questions for decision-makers with very different beliefs and personal backgrounds, and very different business contexts.

Is the Problem Worth It?

At a societal level (and distinguished from what is politically achievable), it is hard to argue that climate risk doesn't justify additional risk management intervention. More societal risk management behavior is clearly 'worth it'.

Companies with little exposure or vulnerability to climate risks under any of the above scenarios, however, face little downside risk from taking a 'wait-and-see' approach regarding climate change and climate policy.

Those companies still may wish to enter the fray on the basis of larger societal and sustainability objectives, or on the basis of perceived economic opportunities.

For companies with greater potential exposure or vulnerability to climate risks under one or more of the scenarios, the question becomes whether that climate risk should be perceived as sufficiently material so as to justify a change in risk management practice. Companies with exposure and vulnerability to climate risks face two timing risks: 1) being 'too late' to respond, resulting in larger than necessary costs and competitive disadvantage; and 2) being 'too early' to respond, which could result in being economically disadvantaged by risk mitigation measures, or in being penalized for near-term mitigation actions when future mandates are implemented.

There will be no 'crystal ball' answer to this question. The reality is that companies can prudently come to radically different conclusions regarding the relative materiality of the climate risks they face. A risk-based reframing of the question can generate useful decision-making insights:

- What is my economic exposure under different policy and market scenarios?

- Which policy scenario(s) is it prudent to be positioned for given my perceptions of their probability and my vulnerability?

- If I better understand climate risks, will I be better able to mitigate those risks vis-à-vis my competition?

- If I better understand climate risks, can I make more robust investment decisions?

Investigating risk scenarios is a better risk avoidance strategy than simply jumping to a perceived 'most likely outcome'. Jumping to a single 'best answer' is clearly susceptible to the 'availability heuristic' and other decision-making biases introduced earlier. Intentionally opening up the corporate risk process to look for 'black swan' risks and outcomes is a good way to avoid being trapped by a single 'best guess' conclusion that unintentionally moves the conversation away from the goals of risk management. This approach to materiality assessment can help quantify the business materiality of the climate risks being faced and help define the nature of the 'long tail' of the risk distribution. It can also help identify opportunities that a company may be able to foster in the near to medium term.

With so many unknowns about how climate change and climate policy will evolve, it is impossible to guarantee that any particular corporate answer to the 'is it worth it' question will result in a measurable and material increase in performance. But if everything were certain, we wouldn't be talking about risk. And it is safe to say that one or more of the scenarios identified above would materially affect the great bulk of the business community. Thus, many companies should answer the 'is it worth it' question in the affirmative.

Can I Succeed?

An individual herring in a large 'herring ball' about to engulfed by a humpback whale in the Arctic Ocean might answer the 'is it worth it' question in the affirmative, and aspire to a risk management strategy. In practice, no matter the magnitude of the risk, the outcome facing that individual herring is likely random. Some will survive, others won't.

If herring slept, the risk management problem faced by the individual herring wouldn't be worth losing any sleep over.

Many companies and organizations are in a similar situation when it comes to climate change risks. These companies may still wish to enter the climate risk arena on the basis of larger societal and sustainability objectives, but not because they can hope to materially influence the risks they face. That option is limited to companies that have the ability to influence their distribution of risk. These are the companies that need to ask 'can I succeed'. For these companies, what constitutes a prudent response to the risks they face under the range of possible future scenarios? Given the political challenges to climate policy, how far can companies go in advancing a risk management agenda? Risk management is rarely free, and shareholders and other stakeholders will be watching.

If the goal were to eliminate or fully hedge potential climate risks, this would be a challenging objective. More realistic for most companies

....

FIGURE 8. Building a tail-risk management strategy.

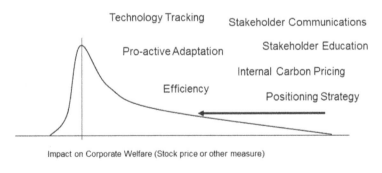

SOURCE: The authors.

....

is to think about how to shorten the 'long tail' of their risk distribution, as illustrated in Figure 8. This approach to managing climate risks addresses the unique characteristics and scale of the climate change problem and can effectively position companies to protect or enhance their competitive advantage over the near, mid and long term. It also allows risk management strategies to be dynamic, being revisited over time as better information becomes available and as economic and political contexts change. There is nothing 'fixed' about corporate climate risks. In most cases, however, competitors choosing the simpler 'wait and see' path when it comes to climate change will be less prepared to manage risks or opportunities they encounter along the way.

....

CHAPTER 8

Managing Corporate Climate Risk

Business has to prepare for climate risk, and there is no question that business climate risk can substantially be managed or positioned for. A variety of measures can contribute to companies' efforts to shift their climate risk probabilities. The first is to make sure one has asked the right questions, whether as part of the 'is it worth it' or the 'can I succeed' processes:

- Am I making technology decisions in the right way?

- Will I win or lose in a carbon-constrained world?

- What are the risks of acting too early or too late?

- Will I be regulated? When? What for? How much will compliance cost?

- Do I understand my options and abatement cost curves?

- Do I face brand, customer, or other stakeholder risk? How much risk is there in long-term capital deployment?

- Will I be competitively disadvantaged?

- Can I create competitive advantage for my company?

Direct Corporate Actions

The following measures show different strategies that companies have used:

- GHG inventories (voluntary or compliance oriented).

- Investigation and pursuance of (or holding in reserve) internal emissions reductions (e.g. energy efficiency, renewable energy).

- Implementation of a regulatory risk assessment and tracking system, including for future carbon prices.

- Internalize a carbon price in investment decisions, in the base case financial analysis or as a sensitivity case.

- Pursue carbon offsets for brand development or regulatory risk hedging.

- Pursue carbon neutrality at the product or corporate levels.

- Take on voluntary emissions targets.

- Educate employees and consumers.

Climate Risk Disclosure

Climate risk transparency, from clarity in the information contained in basic emission inventories, to documenting the basis for environmental and climate risk claims, has been high on investors' and NGOs' climate wish lists. Considerable improvement in this arena has occurred since the first corporate emissions inventories by companies like Stonyfield Farm and Nike more than 15 years ago and the first offset project undertaken in Guatemala by AES Corp. more than 20 years ago. Today, at least some form of emissions and strategy disclosure is routine.[33]

So now that so much information is available, what's next? While a great deal has been achieved with respect to the transparency and reliability of corporate GHG disclosures, disclosures do not themselves deliver environmental benefits. Measuring and monitoring emissions is only a first step to understanding how companies may be affected by climate change and climate change policy, or toward the implementation of active emissions mitigation efforts. Considerable work remains to be done in coming up with a standardized approach to climate risk assessment, and making use of sufficiently standardized assumptions so that the results can be compared between companies.[34]

In considering all kinds of climate risk disclosures, companies will need to be cognizant of whether disclosure mitigates some risks (brand) and increases others (litigation, liability).

Climate Risk Positioning Strategy

A climate risk positioning strategy can help a company adopt a proactive and 'ready-to-respond' approach to changing climate risk context, particularly for risks that can't be materially or cost-effectively mitigated today, whether due to technical or policy uncertainties. Companies with an effective positioning strategy will be able to move much more quickly than their competitors when uncertainties around key variables are narrowed. As illustrated in Figure 9, an effective positioning strategy can:

- Reduce up-front capital outlays by giving a company more confidence in its ability to move quickly as the policy and market environment changes;

- Reduce long-term mitigation costs by taking advantage of early entry business and market opportunities as soon as uncertainty is narrowed to a pre-specified level;

- Demonstrate to shareholders and stakeholders that the company recognizes and is addressing the business and regulatory risks arising from climate change, even if not through public pronouncements;

- Enhance the company's adaptability by thinking through response strategies in advance, allowing for fast and appropriate responses to climate change, regulatory, and GHG market developments as they occur.

....

FIGURE 9. Climate strategy positioning continuum.

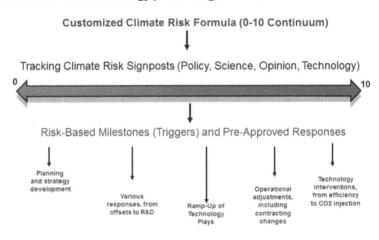

SOURCE: The authors.

....

Going through a climate risk positioning process does not guarantee any particular outcome, and good positioning will mean different things for different companies. Climate risk positioning strategies can make companies 'response-ready' for climate risks that will evolve, or that can't be materially or cost-effectively mitigated today. Companies with effective positioning strategies will be able to move more quickly than competitors when uncertainties around key variables are narrowed, and enhance their competitive advantage.

Encouraging Societal Risk Management

As previously discussed, it is hard to argue that societal climate risk doesn't justify more societal risk management interventions than we have seen to date. But as long as the economic externality of GHG emissions remains, and emissions reduction mandates do not manage societal climate risk, it is unrealistic to expect the private sector to somehow 'solve' the climate change problem.

The nature of the climate change problem, however, is such that a great deal of business climate risk likely falls into the same category as the risk facing a herring ball in the Arctic Ocean. For better or worse, business is along for the ride.

How prudent a strategy is this? If the herring ball could think, wouldn't it make sense to try and avoid the humpback whales? Can business take indirect action to hedge the aspects of business climate risk that it can't directly control? It is clear, for example, that the private sector can have substantial influence when it comes to overcoming a number of the barriers facing efforts to deploy climate change policy.

Given the potential magnitude of climate risk faced by the business community, a strong case can be made that the single most significant risk management measure that could be advanced by the business community is to help overcome the stalemate that has prevailed when it comes to risk-based climate change policy at the societal level. From a business perspective, there are two key questions:

- Are we as business better off incurring the costs of successful climate policy now, in order to avoid the risks of climate change and more severe climate change policy later?; or

- Are we as business better off avoiding the costs of near-term policy, and accepting the risks of climate change and more severe climate change policy later?

These questions raise broad-ranging questions of social and inter-generational equity and corporate responsibility, the answers to which might even differ at different increments along a climate change continuum. For example, how would business analysis of these questions differ when considering 1°C vs 6°C of global temperature change?

For many of the same reasons that public policy has not succeeded in addressing climate change risk, the business community has not tackled these questions directly. Is engaging in this exercise the next stage in the evolution of business climate risk management? Should it be?

....

CHAPTER 9

Conclusions: Reframing Our Approach to Climate Risk

THE POLITICIZATION OF CLIMATE CHANGE and climate risk, coupled with cognitive imperatives like our need to search out patterns in almost anything we observe, has led to a public discourse that is dominated by:

- Arguments and debates over what exactly is already happening, and what exactly we should expect to see in the future, rather than taking a probabilistic view of science-based future outcomes.

- Arguments as to whether climate change is actually occurring, rather than discussing how much climate change would be 'too much', and taking steps to eliminate at least the far end of the risk distribution.

- The notion that uncertainty calls for a discounting of climate risk in policy discussions, rather than an amplification of risk.

This approach to climate change policy development may be fully understandable on the basis of human decision-making, but it is contrary to whole idea of risk-based analysis. It also complicates the nature of private sector risk assessment and management.

It is hard to minimize the potential business importance of the enormous risk gap between today's GHG emissions trends and the 70%+ reduction in emissions called for by scientists just to stabilize the atmospheric concentration of CO_2. This gap inevitably creates a volatile risk management situation for the business community; the wider the gap becomes, and the longer it persists, the greater the risk posed by each of the five risk categories introduced in this report – much as earthquake risk becomes more significant the longer tectonic stresses are building up.

Climate quakes?

Stresses are building in the climate system, stresses that current climate science suggest will inevitably will be relieved through physical or policy 'climate quakes', much as earthquakes relieve tectonic stresses. Climate quakes have the potential to dramatically affect long-term business supply chains, operations, infrastructure, and even employee health and productivity. Based on the science as we understand it today, we have to accept that if we don't relieve the climate stress represented by rising GHG concentrations in the atmosphere, major climate quakes are as inevitable as major earthquakes.

Like the constant earth tremors around the world, climate-related stress is already showing up around the world as 'climate tremors', from melting glaciers to shifting species distributions and pine beetle outbreaks. Some climate tremors are less easy to spot than others, as natural variability in the climate system can obscure them. But scientists tell us that more destructive climate quakes are on the way in the form of changing water regimes, more severe extreme events, and damaging ocean acidification, among many others.

We are also seeing climate 'policy tremors', such as renewable portfolio standards and CAFE standards – small manifestations of climate-directed policy actions. As climate change manifests itself in increasingly obvious ways, climate policy tremors will evolve into climate policy quakes, resulting in greater disruptions and increased business risk. Dramatic emissions reduction mandates, or significant price on carbon, are two potential policy quakes.

Like earthquakes, it's hard to predict precisely when we'll see major climate quakes occur, even if we can confidently anticipate them. It could be next year, or it could be 20 years from now. But like earthquake preparedness, the lack of certainty should not stop us from taking steps to improve the resiliency of societal and business systems, particularly given the likely correlation between the severity of future climate quakes and business-disruptive policy quakes that will follow as after-shocks. This is the almost inevitable outcome of our reactive rather than pro-active political process.

In the 1983 movie *War Games*, a military war games simulation super-computer named Joshua realizes that, like tic-tac-toe, there can't be winners in a nuclear war. Joshua utters memorable words in connection with the impacts of nuclear warfare: '*A strange game. The only winning move is not to play.*' It is likely that we will wish we had taken Joshua's lesson to heart in addressing climate change. Perhaps the single biggest question facing corporate risk management efforts is the degree to which the business community wants to encourage society to come to a Joshua-like conclusion. The business community may be much better off with near-term and smaller 'policy quakes' that avoid the more severe climate and policy quakes that otherwise seem likely to characterize the future.

The potential for business risk and business opportunity based on climate change and climate change policy is greater now than in the past, and will continue to grow as the gap between climate science and climate policy continues to expand. It is important that corporate risk management strategies keep up.

....

CHAPTER 10

Annotated Bibliography

Climate change has received an enormous amount of literary attention in recent years, despite and perhaps because of the continuing political stalemate over global climate change policy. There are multiple climate change literatures, focusing on the science, the policy, the impacts, the mitigation of, and now the adaptation to, climate change, each large in their own right. There are also large related literatures, including the energy, technology, and sustainability literatures, as well as the risk, communications, and decision-making literatures. All of these literatures are relevant to understanding future climate change and climate policy, and hence climate change risk, both at the societal and business levels. No individual, however, can internalize the insights of more than a small fraction of the available literature. The literature presented in this annotated bibliography is just a small slice of the popular climate literature and is intended only to provide the reader with some starting point suggestions for better understanding the topic of climate risk.

For individuals and companies interested in exploring further various aspects of the business climate change risk topic, a Google, Amazon, or Kindle books search will turn up hundreds of titles and links aimed at the general reader. A large part of the popular literature falls into one of two broad categories, reflecting the political polarization of climate change as a problem.

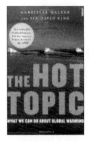 *The Hot Topic: What We Can Do About Global Warming* (2008) represents a genre of 'pro climate action' books that review climate science and climate policy and then transition into a discussion of mitigation and adaptation options and technologies. Books using this general framework are constantly being added to bookstore shelves, taking an advocacy stance on the need for climate change action. Less prevalent overall, but numerous in their own right, are books presenting the 'skeptics' view of climate change. *Red Hot Lies: How Global Warming Alarmists Use Threats, Fraud, and Deception to Keep You Misinformed* (2008) does not focus on the science or policy of climate change, but primarily on the global conspiracy alleged to underpin climate change policy advocacy, and the nefarious motivations of climate change activists.

This annotated bibliography does not focus further on either of these more general literatures, but instead references readily available 'popular literature' source materials that can advance readers' understanding of more specific questions.

- What's happening with the science?

- What about the impacts of climate change?

- What explains the political inaction around climate change?

- What about the role of business in addressing climate change?

- What would it really take to address climate change?

- What does the future hold?

Any literature review of a topic as complicated and future-oriented as climate change should perhaps be prefaced by reference to the forecasting literature itself. *Future Babble: Why Expert Predictions are Next to Worthless and You Can Do Better* (2011) is an insightful and sobering look at the prediction business. Beyond reviewing in some detail many specific predictions, as well as the empirical performance of financial and political gurus, *Future Babble* explores the psychology that leads to the popularity of such predictions, and the trouble it can get us into. The book argues that humans have a deep-seated desire to understand the future. Unfortunately, the predictions we are most likely to listen to – namely unequivocal predictions by forceful and self-confident prognosticators – are precisely the predictions to which we should pay the least attention. While we tend to give more credence to a more detailed prediction or scenario, that, too, has no empirical basis in terms of the likelihood of the prediction coming true.

Future Babble is relevant to thinking about climate risk not because it should be interpreted to suggest that thinking about future climate risk is pointless, but because it should suggest that we should be careful about allowing ourselves to be overly influenced by any particular subjective view of the future. At the end of the day, our understanding of climate risk should be based on the 'Knowns' of climate change, and account for the risks associated with the 'Known Unknowns' and 'Unknown Unknowns' of climate change. Such an approach allows us to paint a probabilistic picture of potential climate change outcomes and climate risk. Most scenarios one can find in the more popular literature fall within this probability distribution of potential outcomes, but we should not be attracted to one or another scenario just because it sounds more convincing or because it fits our individual preconceptions and core beliefs. When delving into the popular literature, keep 'risk' in mind.

Evolving climate science

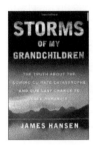

In *Storms of My Grandchildren: The Truth About the Coming Climate Catastrophe and Our Last Chance to Save Humanity* (2009), scientist James Hansen (one of the first scientists to take a public stand on climate change more than 25 years ago) notes that climate change seems to be happening faster than scientists originally predicted and that scientists are inherently conservative and non-alarmist in their approach. The net effect is that the situation is worse than generally recognized. He hypothesizes (although this is not a widely held view among scientists) that at the far end of the risk probability distribution is the possibility that the Earth will pass a tipping point that ultimately would leave the planet looking more like present-day Venus than present-day Earth.

A very different look at climate change is found in the 2011 report *Degrees of Risk: Defining a Risk Management Framework for Climate Security*. It summarizes possible climate change futures using the two axes of climate sensitivity and policy effectiveness, suggesting a worst case of 6–8°C of change by 2100. What is most interesting about *Degrees of Risk* is its argument that if we approached climate change risk management in the same way that we've approached and dealt with other major societal risks, most particularly national security, we would deal with climate change in an entirely different manner.

A subject that has received considerable attention in the popular press is the potential linkage of climate change to extreme weather. In the ebook *Rough Winds: Extreme Weather and Climate Change*, scientist Jim Powell assesses regional weather extremes of 2010 and 2011 in one of the most focused popular looks to date at the subject. He points out, for example, that summer temperatures in Russia in 2010, a summer characterized by massive wildfires, were four standard deviations from the mean. *Climate Safety – In Case of Emergency* (2008) is a purely risk-based look at climate change and climate science. It asks a wide variety of 'what if we're wrong' questions in terms of prevailing wisdom about of the sensitivity of the climate to GHG concentrations, or the possibility of climate tipping points. By looking at climate science and climate change forecasts through this 'what if' lens, *Climate Safety* does a particularly useful job of contributing to popular understanding of the topic.

Anticipating climate impacts

In the 2011 book *The Big Thirst: The Secret Life and Turbulent Future of Water*, Charles Fishman lays out a fascinating history of water science and policy, and asks difficult questions about how climate change will complicate an already strained global situation for the quantity and quality of water available to billions of people around the world. Also in 2011, Paul Epstein and Dan Farber explore the relationship of climate change and human health in *Changing Planet, Changing Health: How the Climate Crisis Threatens Our Health and What We Can Do About It*. Generally speaking, these and many other studies of potential climate impacts paint more serious pictures of climate change than is reflected in other media.

Sources of climate inaction

Instead of focusing on what we need to do about climate change, more literature has begun to focus on the question of why we are not addressing climate change as the science suggests that we should. Two books with different approaches to this question are Michael Hulme's *Why We Disagree About Climate Change: Understanding Controversy, Inaction and Opportunity* (2009), and Naomi Oreskes's *Merchants of Doubt: How a Handful of Scientists Obscured the Truth on Issues from Tobacco Smoke to Global Warming* (2010). Hulme's book explores the many reasons for the protracted debate over climate change, from differing values and fears to the perception and communication of risk, to political maneuvering. Oreskes's book focuses on the 'money' behind the climate change debate, making clear that there is much more than 'the facts' at stake in climate change policy development. *Merchants of Doubt* claims to document an active war that has been waged against climate science for more than a decade by interests with an enormous vested interest in the status quo.

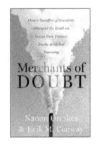

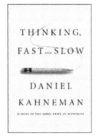

Other important additions to understanding the challenges facing climate policy deal with topics from behavioral economics to communication challenges. In *Thinking Fast and Slow* (2011), Nobel Prize in Economics winner Daniel Kahneman explores decades of research into the functioning of the human mind in decision-making, including risk perception and risk management. The study of behavioral economics, widely attributed to Kahneman, seeks to understand human behavior not as the result of impartial and logical calculations based on the best available information, but as the result of deeply held perception and decision-making biases with their roots deep in human evolution. *Thinking Fast and Slow* can give readers an entirely different perspective on why we make many of the decisions that we do when it comes to complicated subjects like climate change.

As national and international climate policy efforts have struggled forward, it has become increasingly clear that communicating climate change risk poses difficult challenges. One aspect of this is simply the difficulty of communicating complicated messages, a challenge made concrete in Chip and Dan Heath's 2007 book *Made to Stick: Why Some Ideas Survive and Others Die.* This is a fascinating book which has nothing to do with climate or risk, but rather with the unique attributes of human nature that make communication much harder than we might think. The message of *Made to Stick* is relevant to the subject of climate change risk perception, the nature of the public debate over climate change, and the likelihood of future climate policy. The message is supported by Carol Tavris and Elliot Aronson in *Mistakes Were Made*

(But Not by Me): Why We Justify Foolish Beliefs, Bad Decisions, and Hurtful Acts (2008). The authors argue that the psychology of personal beliefs is more complicated than simply presenting someone with 'the facts'; indeed, they argue, fact-based discourse can lead to counter-intuitive results. Problems like climate change face major hurdles in the view of the authors, since aggressive climate policy would challenge the fundamental belief systems of important stakeholder groups, particularly with respect to the need for and scope of government policy required to address the problem. Additional insight into the human psychology of climate change risk perceptions, and to risk perception in general, can be found at Peter Sandman's useful risk communication website (**www.petersandman.com**).

Climate change and business

Many books have tried to build the business case for action on climate change and sustainability, taking one of several approaches:

The 'how-to-manual' approach. These books on climate and sustainability activism (both for the ordinary public and for the business sector) are generally premised on the argument that climate regulation is just around the corner. In *Climate Change: What's Your Business Strategy?* (2008), Andrew Hoffman and John Woody lay out a series of steps that companies should take now to be prepared for the regulatory risk associated with climate change. Four years later, however, this genre is increasingly difficult to relate to the current political climate.

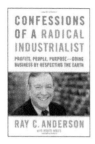

The 'business testimonial' approach. In the 2009 *Confessions of a Radical Industrialist: Profits, People, Purposes – Doing Business by Respecting the Earth,* the late Ray Anderson, one of the earliest corporate advocates of environmental sustainability, argues in a personal way that sustainability makes business sense. He also observes that it's a long and hard battle (but a feasible one, as his success with his company, Interface, demonstrates), and one that requires aggressive public policy to address the economic externality leading to climate change in the first place.

The 'plea to business' approach. A recent addition to the 'climate business' category is reflected in *Climate Capitalism: Capitalism in the Age of Climate Change* (2011) by Hunter Lovins and Boyd Cohen. Their book tries to convince the reader that capitalism itself can solve climate change, and that everyone can make money doing so regardless of their climate change beliefs. In effect, they argue that climate change really isn't that big a problem to solve, while also raising the alarm of potential environmental catastrophe. It's a difficult balancing act.

What will it really take to address climate change?

Rejecting the notion that climate change can be easily solved in today's economic systems, one group of authors suggests that capitalism itself is at the root of the climate change problem. In *The Bridge at the Edge of the World: Capitalism, the Environment, and Crossing From Crisis to*

Sustainability (2008), Gus Speth (one of the founders of the Natural Resources Defense Council, founder of The World Resources Institute, and former Dean of the Yale School of Forestry) argues that conventional environmental activism has failed to make any material difference with climate change, and that climate change is a symptom of the larger problems of capitalism and consumerism. He concludes that we will have to overcome those problems first if we are to get a handle on climate change. This message is consistent with other published challenges to the ability of 'conventional economics' to address problems like climate change, as in the more recent *The End of Growth: Adapting to our New Economic Reality* by Richard Heinberg (2011).

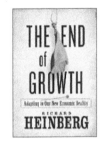

Predicting the future

The Hot Topic as previously introduced was published shortly before Barack Obama was elected President of the United States; at the time, most observers anticipated that adoption of a US national climate change policy was imminent. Countries also expected to adopt a second and stricter second commitment period for the Kyoto Protocol when they convened in Copenhagen in 2009. None of this has come to pass, and as political expectations regarding climate change have changed, the literature has changed as well.

What will actually happen with climate change and our response to it? This too is a growing topic for the climate literature. First is the view

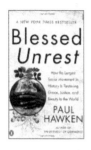

that we will somehow prevail, no matter how many challenges. In *Blessed Unrest: How the Largest Social Movement in History is Restoring Grace, Justice, and Beauty in the World* (2008), Paul Hawken argues that there are approximately 1 million non-governmental organizations existing today with an agenda of promoting environmental and social justice. He argues that these organizations signal the emergence, as yet largely unrecognized, of a massive social movement that will redefine our relationship with the planet and solve social injustice and climate change.

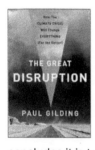

Paul Gilding reaches a different conclusion (yet in some ways similarly optimistic in terms of eventual outcome) in *The Great Disruption: How the Climate Crisis Will Change Everything (for the Better)*. Gilding (a former environmental organizer turned business consultant) argues that humans simply will not take pro-active steps to mitigate climate change. He concludes it is too big a problem and our societal toolbox is not up to the task. He predicts dramatic impacts from climate change in the near future (the book uses the date 2018), impacts that will prove extremely painful and costly in economic and human terms (with the global population falling significantly). He also argues, however, that global community can and will take dramatic and rapid action when the catastrophic implications of climate change are experienced, including a 50% reduction in global emissions in the 20 years after 2018, and a return to 350 ppm of CO_2 in the atmosphere by 2118. *The Great Disruption* can be seen to reflect a growing sense in the climate change and risk management communities that we will not materially act to address climate change until it has become too painful to ignore.

Gilding raises but rejects the possibility of a descent into international chaos as the long-term outcome of climate disaster. Such chaos, however, is the possibility envisioned in Jim Powell's *2084: An Oral History of the Great Warming* (2011). This e-book takes place in the future, and uses the effective approach of interviewing individuals from around the world who were born in or around 2011, and who have lived through 'the great warming' up to 2084. In the interviews, participants discuss how climate change has radically affected their own regions, while also asking the question 'how could this have been allowed to happen?' Written to be scientifically plausible in its description of the future, *2084* is an intriguing story-based approach to communicating climate risk.

Notes and References

1. AES Corp. pursued the first carbon offset project, an agroforestry project in Guatemala in 1989. In 1994 dozens of US companies signed up to the Climate Challenge, a governmentally incentivized program to encourage electric utilities to commit to targets to reduce, avoid, or sequester greenhouse gases by the year 2000. See **http://www.climatevision.gov/climate_challenge/ climatechallenge.html**.

2. Global Risks, 2011. 6th Ed. World Economic Forum. Available at: **http://reports. weforum.org/global-risks-2011/**.

3. United Nations Framework Convention on Climate Change, Article 2. Available at: **http://unfccc.int/resource/docs/convkp/conveng.pdf**.

4. The Intergovernmental Panel on Climate Change was launched in 1988, and released its First Assessment report in 1990.

5. Svante Arrhenius, a Swedish scientist who was awarded the Nobel Prize for chemistry in 1903, is usually credited with being the first scientist to theorize that changes in levels of carbon dioxide in the atmosphere could substantially alter the Earth's surface temperature through amplification of the greenhouse effect. Arrhenius, S. 1896. On the influence of carbonic acid in the air upon the temperature of the ground. *Philosophical Magazine* (Volume 41): 237–276. See also **http://www.aip.org/history/climate/co2.htm**.

6. The natural greenhouse effect can be defined as the trapping of heat in the atmosphere near the Earth's surface. Some of the heat flowing back toward space from the Earth's surface is absorbed by water vapor, carbon dioxide, ozone, and several other gases in the atmosphere and then re-radiated back toward the Earth's surface. Without this trapping of heat, the Earth's surface would be approximately 30°C cooler (54°F), and human life would not be possible.

7. For a definition of radiative forcing and other climate-related terms, see US EPA, Glossary of Climate Change Terms, http://epa.gov/climatechange/glossary.html#R.

8. US Global Change Research Program, http://www.globalchange.gov/publications/reports/scientific-assessments/us-impacts/full-report/global-climate-change.

9. The six gases are carbon dioxide (CO_2), methane (CH_4), nitrous oxide (N_2O), sulfur hexafluoride (SF_6), hydrofluorocarbons (HFCs), and perfluorocarbons (PFCs). See e.g. NOAA. 2011. *The NOAA Annual Greenhouse Gas Index* (Boulder, CO: NOAA Earth System Research Lab), http://www.esrl.noaa.gov/gmd/aggi/.

10. NOAA. 2008. State of the science fact sheet: Ocean acidification (US Dept of Commerce). Available at: http://nrc.noaa.gov/plans_docs/2008/Ocean_AcidificationFINAL.pdf.

11. Intergovernmental Panel on Climate Change. 2007. Fourth Assessment Report, Working Group III, Chapter 3; Hawkins, R., et al. 2008. *Climate Safety* (Machynlleth, Wales: Public Interest Research Centre).

12. Mabey, N., et al. 2011. *Degrees of Risk: Defining a Risk Management Framework for Climate Security* (London: Third Generation Environmentalism Ltd).

13. NASA. 2012. NASA finds thickest parts of Arctic Ice Cap melting faster, http://www.nasa.gov/topics/earth/features/thick-melt.html (accessed 29 February 2012).

14. Phillips, B., et al. 2011. Working paper: Breaking the climate deadlock: Developing a broad and effective portfolio of technology options.

15. Hansen, J. 2009. *Storms of My Grandchildren* (New York: Bloomsbury).

16. Intergovernmental Panel on Climate Change. 2001. Third Assessment Report: Climate Change; Intergovermental Panel on Climate Change. 2007. Fourth Assessment Report: Climate Change. Available at: http://www.ipcc.ch/publications_and_data/publications_and_data_reports.shtml.

17. Stern, N., et al. 2006. *The Economics of Climate Change* (Cambridge: Cambridge University Press). See Wikipedia – The Stern Review for a summary of the report's conclusions, positive and negative criticisms of the report, and issues surrounding the Stern Review's use of a low discount rate: http://en.wikipedia. org/wiki/Stern_Review#Summary_of_the_Review.27s_main_conclusions.

18. Weitzman, M. 2009. On modeling and interpreting the economics of catastrophic climate change. *Review of Economics and Statistics* (Volume XCI, Issue 1): 1–19.

19. Repetto, R. and Easton, R. 2009. Climate change and damage from extreme weather events. University of Massachusetts Amherst, Political Economy Research Institute Working Papers, pp. 9–12. Available at: http://scholarworks. umass.edu/cgi/viewcontent.cgi?article=1176&context=peri_workingpapers.

20. Ibid.

21. IPCC. 2012. *Managing the Risks of Extreme Events and Disasters to Advance Climate Change Adaptation – Special Report of the Intergovernmental Panel on Climate Change* (New York: Cambridge University Press). Available at: http:// www.ipcc-wg2.gov/SREX/images/uploads/SREX-All_FINAL.pdf.

22. Mabey et al. (2011).

23. Ibid.

24. Kahneman, D., 2011. *Thinking Fast and Slow* (New York: Farrar, Straus, and Giroux).

25. Sharot, T. 2011. *The Optimism Bias: A Tour of the Irrationally Positive Brain* (New York: Random House).

26. Shermer, M. 2008. Patternicity: Finding meaningful patterns in meaningless noise. *Scientific American*, 25 November.

27. Gardner, D. 2010. *Future Babble: Why Expert Predictions Fail and Why We Believe Them Anyway* (New York: Penguin Group).

28. See Taleb, N. 2007. *The Black Swan: The Impact of the Highly Improbable* (New York: Random House).

29. See note 1.

30. See note 2.

31. Kahneman (2011).

32. Patterson, K., et al. 2008. *Influencer: The Power to Change Anything* (Provo, UT: Vital Smarts, LLC).

33. See e.g reports of the Carbon Disclosure Project at **https://www.cdproject.net.**

34. Trexler, M., 2012. Climate materiality – the next step in GHG reporting and transparency. In Jackson, F., *Corporate Climate Risk Disclosure – An Executive Guide to Reporting and Disclosing Climate Risk* (London: Environmental Finance Publications).

....

Milton Keynes UK
Ingram Content Group UK Ltd.
UKHW040712141024
449569UK00013B/615